AF601290

Chasing the Mongolian DEATH WORM

Anna Anderhagen

Big Buddy Books
An Imprint of Abdo Publishing
abdobooks.com

abdobooks.com

Published by Abdo Publishing, a division of ABDO, PO Box 398166, Minneapolis, Minnesota 55439.

Printed in the United States of America, North Mankato, Minnesota
102023
012024

Design: Denise Hamernik, Mighty Media, Inc.
Production: Mighty Media, Inc.
Editor: Liz Salzmann
Cover Photograph: FantasyAssets/Adobe Stock
Interior Photographs: Adobe Stock, p. 11; Bettmann/Getty Images, p. 13; Dmitry Fch/Shutterstock Images, p. 27; domnitsky/Adobe Stock, p. 19; Eleanor Scriven/Shutterstock Images, p. 15; Jaime Chirinos/Science Source, p. 9; kuco/Adobe Stock, p. 29; Luis Felipe Carvalho de Lima/Adobe Stock, p. 23; Picasa/Adobe Stock, p. 17; Pieter0024/Wikimedia Commons, p. 5; Stu Porter/Shutterstock Images, p. 25; Tokareva Irina/Shutterstock Images, p. 7; Tomas Janousek/Adobe Stock, p. 21
Design Elements: Joko/Adobe Stock (tape and paper); Melica/Adobe Stock (polaroid frame); Net Vector/Shutterstock Images (map); Oleg Iatsun/Shutterstock Images (compass); Piman Khrutmuang/Adobe Stock (paper); STILLFX/Shutterstock Images (background texture); TatianaMakhakhei/Shutterstock Images (footprints); Wandeaw/Shutterstock Images (background texture); Yevhenii/Adobe Stock (tape)

Library of Congress Control Number: 2023939282

Publisher's Cataloging-in-Publication Data
Names: Anderhagen, Anna, author.
Title: Chasing the mongolian death worm / by Anna Anderhagen
Description: Minneapolis, Minnesota : Abdo Publishing, 2024 | Series: Chasing cryptids | Includes online resources and index.
Identifiers: ISBN 9781098291921 (lib. bdg.) | ISBN 9781098278823 (ebook)
Subjects: LCSH: Worms--Juvenile literature. | Monsters--Juvenile literature. | Animals, Mythical--Juvenile literature. | Folklore--Juvenile literature. | Cryptozoology--Juvenile literature.
Classification: DDC 001.944--dc23

CONTENTS

CHAPTER 1

CHASING CRYPTIDS

You are hiking in the desert. You feel the sand move under your feet. An enormous red worm pops its head out of the sand! It raises its upper body and opens its mouth. You see many rows of spiky teeth. Could this be the Mongolian Death Worm cryptid you have been looking for?

Mongolians call the worm *olgoi-khorkhoi*, which means “intestine worm.” They call it that because it looks like the insides of a cow.

CHAPTER 2

WHAT IS A CRYPTID?

A cryptid is an animal that has not been proven to exist. There are stories about many different cryptids. One is the Mongolian Death Worm. It lives under the sand in the Gobi Desert.

Many people think they have seen the Mongolian Death Worm. But no one has been able to prove it.

The Gobi Desert gets most of its rain in June, July, and August. Many people say this is when the Mongolian Death Worm comes to the surface.

CHAPTER 3

SPITTING VENOM

There have been many stories of the Mongolian Death Worm attacking humans. Some people say it spits poisonous **venom**. Others say it can **electrocute** people. The Mongolian Death Worm is feared by many people in Mongolia. Even mentioning its name is bad luck!

People claim that the Mongolian Death Worm's venom kills anything it touches.

RUSSIA
MONGOLIA
GOBI DESERT
MONGOLIAN DEATH WORM SIGHTINGS
CHINA
N
W
E
S

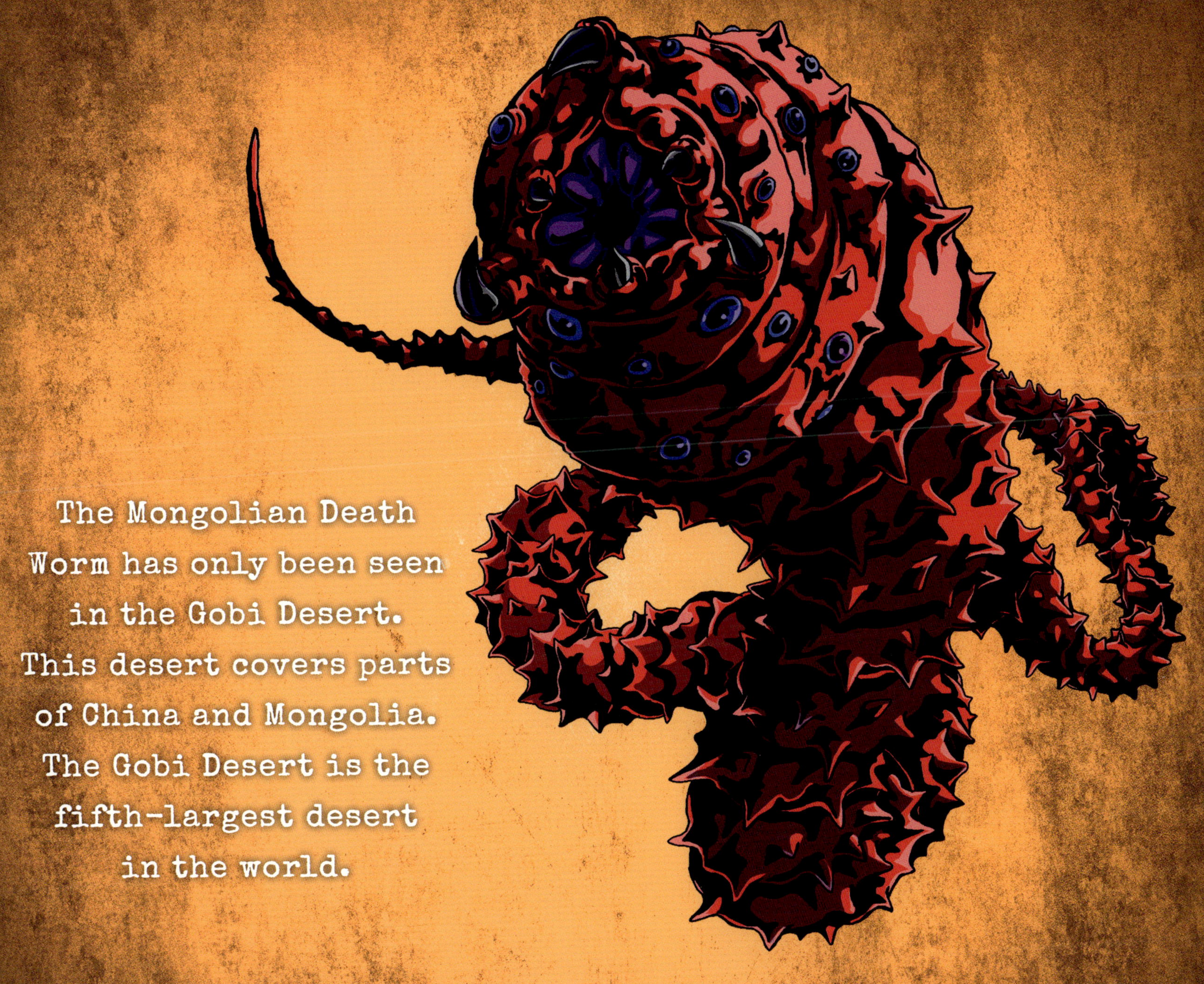

The Mongolian Death Worm has only been seen in the Gobi Desert. This desert covers parts of China and Mongolia. The Gobi Desert is the fifth-largest desert in the world.

CHAPTER 4

GENERATIONS OF STORIES

Zoologist Roy Chapman Andrews was the first to write about the Mongolian Death Worm. Andrews studied wildlife in Mongolia. He heard many stories about the worm from the Mongolian people. They have spoken of the Mongolian Death Worm for more than 100 years.

Andrews (*left*) wrote about the worm in his 1926 book, *On the Trail of Ancient Man.*

CHAPTER 5

EYEWITNESS ACCOUNTS

There are few eyewitness reports of the Mongolian Death Worm. The worm is said to be so deadly that no one has lived to talk about it.

A story from the 1990s tells of a boy playing with a yellow ball. The worm came to the surface. The boy touched the worm and died. Some say the worm is attracted to the color yellow.

Very few people live in the Gobi Desert. Those who do mostly work raising livestock.

CHAPTER 6

TRACKING THE WORM

Zoologist Ivan Mackerle traveled to Mongolia three times to search for the worm. He went in 1990, 1992, and 2004. Mackerle listened to people who claimed they knew someone who saw the worm. Many **described** the worm as dark red. It was as thick as a man's arm.

Mackerle believed the worm eats Goyo flowers that bloom in the summer.

Mackerle thought loud sounds might make the worm come to the surface. So, he made a thumping machine to send sound **vibrations** under the sand. He hoped the sounds would make the worm surface. However, Mackerle never saw the Mongolian Death Worm.

Luring worms with vibrations is called worm charming. The vibrations are like a predator's movements. So, the worms flee to the surface.

CHAPTER 7

DESERT EXPLOSIVES

In 2009, David Farrier and Christie Douglas thought they could blast the worm out of the sand. They set off **explosives** in the Gobi Desert. They thought the Mongolian Death Worm would come to the surface when it felt the blasts. But they did not see the Mongolian Death Worm.

Most of the Gobi Desert is rocky. Only 5 percent of it is covered with sand dunes.

CHAPTER 8

CRYPTID KEEPER: RICHARD FREEMAN

Richard Freeman is a **cryptozoologist** who **specializes** in reptiles. He has studied more than 400 different species. Freeman doesn't believe the Mongolian Death Worm is a worm. He thinks it is a giant member of the reptile group *amphisbaenas*. This is a kind of wormlike lizard that **burrows** in the ground.

Members of the *amphisbaenas* group have strong skulls. They use them to dig underground.

CHAPTER 9

STUDYING THE WORM

Some scientists believe the Mongolian Death Worm is a new snake species. They think it is a member of the cobra family. Many cobra snakes can spray **venom**. Also, cobras raise their heads and upper bodies before they strike. This is similar to many of the **descriptions** of the Mongolian Death Worm.

Cobras can spit venom up to 8 feet (2.4 m) away.

CHAPTER 10

WORM OR BOA?

Scientist Yuri Gorelov thinks the Mongolian Death Worm is a Tartar sand boa. This snake lives in dry, sandy **habitats**. Sand boas bury themselves in the sand just like the Mongolian Death Worm. Gorelov showed a Tartar sand boa **specimen** to people who lived in the Gobi Desert. They said it looked like the Mongolian Death Worm!

Tartar sand boas are among the longest sand boas. They can be up to 4 feet (1.2 m) long.

CHAPTER 11

DOES IT EXIST?

Hundreds of people have told stories of the Mongolian Death Worm. There are many sketches of it. But there are no photos or videos. No one can prove that the worm is real.

Most scientists do not think the Mongolian Death Worm exists. If it did, there would be more **evidence**. They suggest that people are actually seeing a snake or lizard. What do you think?

NAME: Mongolian Death Worm

CLASSIFICATION: worm

COUNTRIES: Mongolia, China

HABITAT: desert

DESCRIPTION:
- 3 feet (0.9 m) long
- dark red
- spiky teeth
- as thick as a man's arm

PROVEN TO EXIST: not yet

burrow—to dig a tunnel or a hole in the ground.

cryptozoologist—someone who looks for and studies legendary animals to determine whether they exist.

describe—to tell about something with words. Such a telling is a description.

electrocute—to injure or kill with an electric shock.

evidence—facts that prove something is true.

explosive—a substance used to blow up something.

habitat—a place where a living thing is naturally found.

specialize—to pursue one branch of study.

specimen—an individual, item, or part that is typical of a group or a whole.

venom—a poison made by some animals and insects.

vibration—a tiny back and forth movement.

zoologist—a scientist who studies animals and their behavior.

To learn more about the Mongolian Death Worm, please visit **abdobooklinks.com** or scan this QR code. These links are routinely monitored and updated to provide the most current information available.

INDEX